CHAPTER 1

THE ALARM BELL RANG. I WOKE UP WITH JOY AT THE FACT THAT TODAY WAS FRIDAY. ONE MORE DAY OF HARD-WORKING, BORING SCHOOL! I GOT UP FROM MY BED, GOT DRESSED, HAD BREAKFAST, AND LEFT 30 MINUTES APART FROM MY DAILY ROUTINE. THIS IS BECAUSE I COULD'NT HELP BUT NOTICE THAT MY FATHER, JORGE WINTER, WAS GETTING IN THE CAR WITH HIS PJS ON. I GOT IN MY MERCEDES-BENZ AND FOLLOWED AFTER HIM.

It wasn't long before Dad reached his destination: NASA Space Center. Wait, my dad didn't have any work today. I snuck in after him. My dad was the CEO of NASA, so I perfectly understand if he has to work, but at 2:00 PM?

Turns out he was there for a meeting. There were armed guards at each of the doors. "Welcome, Mr. Winter," the attendant said. Dad said, "What am I here for?" The attendant, whose name was Sophie Marshall, who looked

NOTHING OVER 20, SAID, " YOU HAVE BEEN RECRUITED FOR A SPACE MISSION THAT TRAVELS AROUND THE SOLAR SYSTEM, COLLECTING DATA. YOU ARE THE SPACE COMMANDER, AND YOUR MEETING IS INSIDE. I TAKE IT YOU DON'T WANT TO INVITE LITTLE ANDREW?" I SUDDENLY CAME OUT OF HIDING. I CRIED WITH ANGER, "YOU TWO SONS OF.. WHAT'S GOING ON HERE?" BOTH GUARDS POINTED THEIR MUSKET-LOOKING RIFLES AT ME. "YOUR FATHER HAS BEEN RECRUITED FOR A DADNGEROUS AND LONG

SPACE JOURNEY, AND DOES NOT SEE YOU GOING." I SAID, "WHAT THE HECK, DAD? I THOUGHT WE WERE BONDING FAMILY FOREVER. DON'T DITCH ME LIKE THIS." HE REPLIED, "IT'S FOR YOUR OWN GOOD, SON. I MEAN, DO YOU THINK A RECENTLY-FINISHED GROWING, SIX-FOOT-FIVE ADULT LIKE YOU, ESPECIALLY YOU, WHO HAS NOT GOTTEN OFF HIS STUPID GAME SYSTEM ALL DAY AND THEN GOES TO HIS "CLUBS" WITH A CREDIT CARD? I'M PRETTY SURE THE ANSWER WOULD BE NO. BON VOYAGE. LOVE YOU!" THEN, HE

WENT INSIDE WITH THE OTHER SELECTED FOUR. I, ON THE OTHER HAND, WAS SHOCKED. IT WAS LIKE MY DAD WAS THE STRESS MONSTER, AND I WAS THE ONE TO CUT HIS STRESS CORD AND RELEASE ALL THE STRESS. BUT, INSTEAD OF ENDING THE CONVERSATION, I JUST COMMENCED ANOTHER ONE. I LEFT UNHESTITANTLY ON MY MERCEDES AND HEADED TO SCHOOL. BOY, DID I HAVE A SHOW AND TELL TODAY.

CHAPTER 2

IF ANYONE OTHER THAN CHILDREN MIUNDERSTOOD THE CODE WORD "GOING TO SCHOOL", IT MEANS THE TOTAL OPPOSITE. I WAS HEADED TO CHERRY WASHINGTON'S HOUSE TO GET SUPPLIES. I MEAN, WHO CARES IF I'M MISSING ALGEBRA II OR SOME SHIT? I RANG THE DOORBELL. CHERRY WAS A FINE WOMAN, ABOUT 24 YEARS AT THE MOST. SHE HAD BOUGHT herself a Bruce Wayne looking manor. Look out, here comes the next Batwoman! Anyways, Cherry was the hot looking babe senior year. Spanking that juicy fat behind of hers would be an honor. Senior year, I tried to ask her out during the winter dance, and she spilled the entire vodka spiked punch bowl on me. You can see why I'm the laughingstock of the entire school and she's the smoking hot diva. I've been barking up that tree for

YEARS, PEOPLE. YEARS. SHE OPENED THE DOOR. "WHAT DO YOU WANT, ANDREW?" I RESPONDED AS FAST AS THE FLASH IN A WHEELCHAIR. "UHH..HUH.....UMMM..I COULD USE SOME DUCT TAPE AND A KISS." SHE RESPONDED, "I'LL GIVE YOU THE DUCT TAPE, I GOT PLENTY, BUT NOT MY HAND IN MARRIAGE! ANDREW, WE'VE BEEN THROUGH THIS ALREADY. YOU USED TO BE MY BOYFRIEND, BUT IT JUST DIDN'T WORK OUT. DON'T WORRY. WE CAN STILL BE FRIENDS." I NODDED, AND LEFT THE LANDFILL OF A HOUSE SHE lived in. I was headed toward NASA. I was going to force that son of a gun Marshall to give me the name and location of the ship, or else.

The parking lot was full of tourists and other people strolling around with their families. I scanned my security card and entered the "<u>employees only</u>" section. I snuck up behind Marshall, who was checking her emails, and tied her up from behind. I said "give me the name of the ship and

ITS LOCATION, OR PAY THE DIRE CONSEQUENCES!" SHE SAID, 'WHAT CONSEQUENCES? AS FAR AS I REMEMBER, I HAVE MORE CLEARANCE THAN YOU, SO WHY DON'T YOU GO SCREW YOURSELF IN THAT CORNER? GO AHEAD, I'M WAITING." THAT REALLY MADE ME RAGE. "THEY LEFT DIDN'T THEY?" I SAID. SHE NODDED, AND TOLD ME EVERYTHING ONCE I TOLD HER I WANTED TO EXCHANGE WORDS WITH MY DAD. TURNS OUT THEIR SHIP IS CALLED TRANSNAVIGATOR 13, AND I

WAS GOING TO BE THE ONE TO
HIJACK THAT SHIP.

CHAPTER 3

The airfield was vast, but
stilled filled with waiting
people. They were there to
witness the second launching
of the Solar System spaceship,
Jupiter 16, the one I was
hitchhiking. The people to me
were like obstacles at the time
of rush hour and I was the
madman going over the speed
limit to get to my crappy,
boring, but yet moderate-

paying job. There weren't any guards at the posts like in Sophia Marshall's office. Thank goodness! The broom closet was the first hiding place I came to. I hid behind all the dusty brooms and waited.

It's been 3 hours since I went in that closet, and we haven't launched. Or have we? Suddenly I felt something at the door. Then, a girl broke it down. "Who are you and how did you get on this ship?", she said.

"My name's Andrew. What's yours?" She said, "Heather. Heather Larson. That brings us to question number 2. How and why did you get on this ship?" To that question I replied, "My dad is on the other solar system explorer whatchamacallit. So, I hitchhiked a ride here to crash and follow him and make sure he makes it back."

She took the excuse, and let me crash with her. SHE was the sexiest girl I had ever seen. Forget about Cherry. This girl is TWICE as much as Cherry will ever be. She also had finer characteristics that Cherry could never have. Proper etiquette at the dinner table, constant uses of the word please and thank you, and good-night kisses. **GOOD-NIGHT KISSES.** Cherry always told me to start meeting new people, and that has failed. Until now. As we slept, the hovering ship was carried majestically over the star- filled horizon.

Chapter 4

It was 3:00 AM, or so I think it was, when a high pitched, deafening sound broke the silence. The window near me cracked, and sharp objects started getting sucked out of the dormitory. Heather and I were, at this point, shocked. We were hanging on for dear life, hoping that one of us didn't get pierced and critically injured by any sharp objects, or get sucked out. Without hesitation, we both made a run for it when the time came, dodging any lethal devices that might be hurtling our way. We were running to the escape pods to get to safety, and I tripped. "Oh, man! I'm going to die. Please try not to let go!" She pulled me back up and in the distance, I caught a glimpse of two yellowish-greenish colored critters, ripping the other crew members apart. We got into the escape pod and pressed *RELEASE*. There was a countdown, and we put on our helmets and oxygen tanks. It was like snorkeling in the Great Lakes. Only this was a totally different situation. Boy, do I miss home. 3, 2, 1, LAUNCH! We were catapulted into the star glittered horizon, and Heather took control of the pod. Absolutely jam-packed with fear, I said, "What in the world was that back there?" She responded, also with fear in her eyes, "That, my friend, was a bona fide alien attack, and you and I are the only survivors." I looked back at the remains of the dead ship one more time, and then Heather put the mode on *HYPERSPEED.*

After a few good hours, the *HYPERSPEED* mode had stopped and Heather woke me up. We were at a safe house in Mars' safe location, Acidalia Planitia, and took a good rest. The next day on Mars for us would be known as Sol 2. Yet, we'd only stay 2 more sols before heading to the NASA space station. My dad must be dead worried about me. He must be scavenging the entire Solar System by himself looking for his son, Andrew. Oh well, Dad, same here.

Chapter 5

After two sols finally passed, we were ready to leave Mars to the nearest space station. We got into the escape pod, but it only had ten percent left. We flew out of Acidalia Planitia, and we were on our way. As we passed the many planets of the Solar System, I got to know

Heather better. Her home was Michigan too, and she was a sophomore too. The only problem was that she went to Ann-Arbor for a bachelor's degree, and I went to Michigan State University on a fully-paid basketball scholarship. Heather had a hard childhood growing up, and as a result, left her family for the military at 18 and was a survivor. She then went to Ann Arbor and sophomore year, she was selected for this solar system mission at NASA, and accepted the offer.

We passed by Jupiter, the former landing place for Jupiter 16. A larger and much more advanced ship passed by us, their headlights on as if they were looking for something. I caught a glimpse of the ship's name. It was my dad's ship, Transnavigator 13. I said to Heather, "Heather, can you trust me?" She answered that question correctly. "I...I... I guess so." I told her to wave to the mother ship since it was my father's. I was coming home.

Chapter 6

As my dad caught a glimpse of me, he ran up to me and gave me a bear hug. He said, "Son, are you all right? You scared me to death!" I erupted into tears and began to cry on my dad's shoulder. I said, hesitantly, with tears, "I'm...I'm sorry Dad. I was an imbecile and a jerk before, but my ways have changed. I'm sorry, Dad!" After that drama, he didn't respond. He turned to Heather. He said, "What are you doing here?" To that, she responded, "Heather Larson,

NASA space unit. Our ship was attacked by an extraterrestrial entity. No survivors." My dad looked shocked. Suddenly, there was a static buzzing on the television.

The television blurred out a creepy alien looking creature, and he said, "Hello fellow humans. My name is king Sahara Tolkien, king of Arda. I see the survivors of my attack. Now I know."

My Dad struck back, saying, "What do you want? You already almost single-handedly murdered my son and this young lady! What more can you possibly be planning?"

"If you must know, I won't stop until I have complete control over the Earth. And, Mr. Winter, what I'm planning is a civil war, and you will be the targets."

At that instance, the doors flung wide open and gravity started sucking everything out of existence. With Heather holding on to me, and Dad hyperventilating, there was nothing we could do. As we fell out, I thought about what King Tolkien said. If he wants a civil war, he's going to get one.

www.ingramcontent.com/pod-product-compliance
Lightning Source LLC
Chambersburg PA
CBHW021450170526
45164CB00001B/466